The Moving Continents

BY FREDERIC GOLDEN

The Moving Continents

DRAWINGS BY INGRID NICCOLL

CHARLES SCRIBNER'S SONS
NEW YORK

551.4
G618m

TO DAVID AND MITCHELL

who believed in continental drift
from the start

ACKNOWLEDGMENTS

The exciting new theories of the earth are the product of the work of many researchers, too numerous to cite here. But I would like to acknowledge a particular debt to scientists who have given generously of their time and advice in the preparation of this book: Dr. Robert S. Dietz, Dr. Maurice Ewing, and Dr. Lynn R. Sykes, all of whom are important to the story of continental drift, and Dr. Irwin I. Shapiro, professor of physics at the Massachusetts Institute of Technology. I wish also to thank my wife, Reva Betty Golden, for her extremely helpful advice and assistance.

Acknowledgment is also made to the following for permission to reproduce the illustrations on the pages indicated:

Page 16: Edwin H. Colbert, *The Age of Reptiles,*

CONTENTS

ILLUSTRATIONS

Drawings and diagrams

Photographs

The Moving Continents

INTRODUCTION

LATE in 1969, during an expedition to the frigid wastes of Antarctica, a team of American scientists made a remarkable discovery. They had barely finished setting up their base in the Transantarctic Mountains at the edge of the Ross Ice Shelf when they found the fossilized skull of a long-extinct reptile, the *Lystrosaurus,* kin of the dinosaur.

From similar fossils that had been recovered elsewhere in the world, the scientists knew that such small, hippopotamus-shaped creatures had dwelt long ago in the steamy, tropical swamps of Africa and India. How, then, did this particular Lystrosaurus make its way to a cold, snow-clad region only 400 miles from the South Pole?

An artist's conception of Lystrosaurus

In its day, more than 200 million years ago, the two-foot-long Lystrosaurus was probably a reasonably good swimmer. But it was hardly powerful enough to have paddled across many hundreds of miles of open sea from India or Africa to Antarctica. Besides, the little reptile was a fresh-water animal and could not, in any event, have survived very long in the salty oceans. So, since it was clear that Lystrosaurus had not made such an extraordinary journey by itself, the scientists had to look for another explanation for its unlikely presence at the bottom of the world.

There was only one other possibility—that Antarctica itself had moved. During its long trip, the scientists concluded, the wandering continent had simply carried the body of the ancient reptile along with it.

A scientific absurdity? Not in the eyes of a few daring scientists. For many years they have seriously supported the idea that the continents have actually floated across the face of the globe like giant ice floes. The German scientist-explorer Alfred Wegener (1880-1930) had actually lost his life trying to prove this. According to Wegener, the continents—North and South America, Eurasia (Europe and Asia), Africa, Australia, and Antarctica—were once a single supercontinent, millions of years before man made his appearance on earth. Then, for mysterious reasons, this huge land mass began breaking up into smaller pieces. Over the ages, these fragments, which are today's continents, gradually drifted apart until they reached their present positions. Moreover, Wegener said, the fragments are still moving and probably will continue to move indefinitely. But in the opinion of the vast majority of scientists, Wegener's theory of continental drift, as it was called, was so far-fetched that they long ridiculed it as an outright scientific fantasy.

Surprisingly, in the past few years, the fantasy has suddenly begun to look like fact. Not only have the remains of many tropical plants and animals been found in ice-covered Antarctica, but other strong scientific evidence has been turned up elsewhere in the world. Like clues in a mystery story, the evidence points to an inescapable conclusion: that the continents were once located in parts of the earth thousands of miles from where they are now.

At the very least, these findings have made a belated hero of a man once laughed at by many of his colleagues. Much more important, they have thoroughly revolutionized the science of geology. In fact, the current activities of geologists and other earth scientists seem as exciting as those of physicists during the 1920s and 1930s. Just as their discoveries about the basic nature of the atom led to the development of nuclear power, so those of today's earth scientists may pave the way for equally spectacular scientific advances: the tapping of new sources of energy from the earth, the mining of vast new mineral deposits to replenish the dwindling existing supplies, and the forecasting and perhaps controlling of devastating earthquakes.

Astonishingly, this scientific revolution is taking place at a totally unexpected moment in man's history. At the very beginning of his exploration of the far reaches of space, he is only starting to learn that his own restless, changing planet may be as scientifically interesting and challenging as more distant targets in the solar system.

The pages that follow will show how this startling new image of the earth emerged—through the unexpected recent discovery that the continents are, in fact, on the move.

CHAPTER ONE

The Catastrophists Against the Uniformitarians

Picture in your mind's eye the face of the earth: the outlines of the continents, the great expanse of the seas, the long chains of towering mountains. Thanks to the skills of modern mapmakers, these terrestrial features are almost as easily visualized by us as the familiar streets of our cities and neighborhoods. If anyone doubts the accuracy of available maps and globes, he need only look at the superbly detailed photographs of the earth taken by orbiting astronauts. They offer convincing proof of the size, shape, and location of the earth's major landmarks.

Man, of course, did not always have such detailed knowledge of his home planet. It was not until the beginning of the great age of exploration, when such fearless voyagers as Christopher Columbus, Vasco da Gama and Ferdinand Magellan set off on their farflung journeys, that Europeans even began to suspect the existence of great land masses beyond the known shores of Europe, Africa, and Asia. By the sixteenth century, however, the sketchy information gathered on these expeditions was finally assembled by accomplished cartographers, like the Flemish geographer Gerhardus Mercator. For the first time, men had produced recognizable maps of the entire earth.

These pioneering global charts not only showed the vastness of the New World but also revealed another geographic surprise: there was a giant bulge along the eastern coast of the newly discovered continent of South America. And with only a slight stretch of the imagination, it could be made to fit almost perfectly beneath the matching protuberance of the opposite coast of West Africa.

The remarkable resemblance between the Old and New Worlds was noticed as early as 1620 by the English philosopher-scientist Sir Francis Bacon. But he apparently had no suspicions about the origin of this curiosity; which would not have been unexpected. In those days, man's views of nature were still dominated by literal interpretations of the Bible. Indeed, it was often highly dangerous to voice contrary ideas. Bacon's contemporary, the Italian scientist Galileo Galilei, for example, was put on trial for his life as a heretic for insisting that it was the sun—and not the earth—that occupied the center of the universe.

Not surprisingly, the earliest geologists also looked to the Bible for explanations of how the earth's features were formed. They were especially impressed by the story

20

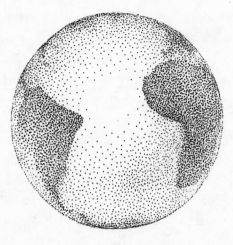

The matching coasts of South America and Africa

of the flood in the Book of Genesis. To punish man for his sins, so·the story goes, God let rains fall on the earth for forty days and forty nights, killing off all living things except Noah, his family, and the animals that had been loaded onto the ark.

Inspired by this poetic account, a whole school of earth scientists emerged. Known as "flood geologists," they held that great catastrophes like the biblical flood could, in a single stroke, reshape vast areas of the earth—building mountains, carving out valleys, and opening up oceans. Their ideas are called the doctrine of catastrophism. Two particularly inventive flood geologists, or catastrophists, were the seventeenth-century French monk François Placet and the eighteenth-century German theologian Theodor Lilienthal. In what are probably the earliest arguments on record for continental break-up, they contended that the biblical flooding was so devastating that it destroyed old continents and created entirely new ones.

21

Considering what was known about the earth in their day, such speculations were not altogether implausible. Using the Scriptures as their calendar, these early geologists reckoned the earth's age as a mere 6,000 years. Thus, terrestrial features such as high peaks or deep canyons had either existed from the Creation or been formed almost overnight shortly thereafter; there was simply too little time available in their calculations of the earth's age for any slower buildup. But as scientifically minded men began looking more carefully at the world around them, they made some perplexing observations.

Wherever enough of the earth's hard underlying bedrock poked through the soft topsoil, they discovered that it was made up of distinctly different strata, or layers. The material had obviously been laid down very slowly, over periods of time much longer than a few thousand years. Even more significantly, the strata were found to be twisted and deformed; but this twisting was much too gentle and gradual to have been caused by any single catastrophe.

Nor could catastrophism account for another strange discovery. In many places, geologists stumbled onto the stony skeletons of ancient plants and animals that were no longer in existence. Still more puzzling, some of the remains were fossils of sea creatures, yet they were found high and dry on the sides of very tall mountains. Could the receding waters of the great flood in the Book of Genesis have left a few stray sea animals at such heights, just as Noah's ark had been stranded on the summit of Mount Ararat? Not likely; the marine fossils were scattered on too many mountains to have been distributed by a single catastrophe.

By then the theory of catastrophism was in trouble.

Although no reliable scientific methods were yet available to determine the age of the ancient fossils or twisted bedrock, two ideas slowly began to dawn on the learned men of the time. First, the earth was apparently much older than the Bible suggested; second, it was changing at a slower-than-catastrophic rate. Yet it took the shrewd insights of an observant Scotsman named James Hutton to realize the full import of the new discoveries.

In 1785, in a memorable series of lectures before the Royal Society of Edinburgh, Hutton expounded his ideas. A lawyer turned physician turned farmer, Hutton knew from his long walks across the countryside that local disasters could occasionally rip up the landscape. A historical example was the devastating eruption of Mount Vesuvius in A.D. 79 which destroyed the flourishing Roman city of Pompeii in southern Italy. But on the whole, Hutton pointed out, the earth's face was being remade at a much more gradual pace. In fact, the changes are usually so slow that they cannot be observed in one human lifetime or even several lifetimes but only over periods of many thousand years. Furthermore, he said, the agents of change were not sudden catastrophes but such slow, steady forces as the pounding of waves against a shoreline, the hammering of rain and wind against hard rock, and the washing away of mud and debris by fast-moving rivers.

For all their brilliance, Hutton's ideas at first baffled his audience. Lacking a gift for language, he was unable to make himself completely understood. Fortunately, his good friend John Playfair, a professor of mathematics and philosophy at Edinburgh, had no such shortcomings. In 1802, five years after Hutton's death, he published a book, *Illustrations of the Huttonian Theory of the Earth,*   23 which carefully and clearly explained his friend's theories, plus some of his own. Together, they became known as the principle of uniformitarianism (after Hutton's central argument: that the earth's surface is being changed uniformly, rather than catastrophically, over great spans of time). In spite of its jawbreaking name, the principle of uniformitarianism was the intellectual spark for the real beginning of modern geology, just as Einstein's theories of relativity later heralded the birth of twentieth-century physics.

Hutton's bold new theories, however, could not explain all geological mysteries. While the slow, grinding action of sea, wind, and rain might be able to sculpture certain surface features, it could hardly reshape whole continents—to say nothing of breaking them up and moving them apart. Consequently, scientists who tried to explain the matching coastlines of Africa and South America still had to resort to earthshaking events, or catastrophism in a slightly updated version. The nineteenth-century German explorer-naturalist Alexander von Humboldt, for instance, thought that the Atlantic Ocean was created early in the earth's history by a current of water that surged across the globe like a huge tidal wave. One reminder of that unwitnessed catastrophe, said Humboldt, is the fit of the opposing shorelines.

Later, in 1858, an American writer, Antonio Snider, surpassed Humboldt by envisioning a whole series of catastrophes. As the young earth cooled and crystallized, he said, most of the material of the future continents gathered into a single clump on one side of the globe. Then, looking like a lopsided basketball, the unbalanced earth developed such internal stresses and strains that the

24

single land mass began to crack and hot lavas welled up through the crevices. Simultaneously, Noah's flood poured down so much rain that the raging waters actually pushed part of the cracked continent to the other side of the earth. Thus, said Snider, the Americas were born and the wobbly earth's equilibrium was restored.

Not to be outdone, George Darwin, son of the famed biologist Charles Darwin, evolved an equally fanciful explanation for the origin of the intriguing coastlines. In 1879, Darwin wrote that the young earth was probably rotating so fast that a huge chunk was torn away, giving rise to the moon. The great hole that was left behind (the Pacific basin) soon filled with molten lava, and the lighter continental material began floating around in it like bits of bread in a bowl of thick, hot soup. To some scientists this sounded plausible. Until it was shown that the Pacific basin could never have held enough material to form a celestial body as large as the moon, Darwin's theory remained a pet explanation for how the continents might have drifted apart.

The real flaw in the efforts of Darwin and other theorists to prove continental drift, however, was much more fundamental. By relying on the thoroughly rejected doctrine of catastrophism, the drift theorists invited the open scorn of their fellow scientists, most of whom were dedicated uniformitarians. To be sure, a few thoughtful scientists tried to bring continental drift within the framework of uniformitarianism. In 1885, for instance, the Austrian geologist Eduard Suess actually published a map showing how the southern continents could be fitted together into a single land mass that he called Gondwana-land (after India's fossil-rich Gondwana province). A

25

generation later, in 1908, an American geologist named F. B. Taylor also proposed arguments in favor of continental drift.

But it was only with the appearance of the dedicated and persevering Alfred Wegener that continental drift acquired any sort of scientific respectability. For the first time, a serious scientist tried to explain drift by modern geological ideas. Rejecting the notion that catastrophes had moved the continents, Wegener amassed piles of evidence to show that continental drift was slow, gradual —and uniform.

CHAPTER TWO

A Man Named Wegener

EVEN as a youth, Alfred Lothar Wegener showed the strong qualities of mind and spirit that characterized him throughout his highly controversial scientific career. A Protestant preacher's son, he probably could have settled down to a comfortable middle-class life as a teacher or merchant in his native Berlin. But by the turn of the century, when he was still in his early twenties, it was clear that he was bent on a more adventurous life. In his spare time he was always taking long hikes, climbing dangerous mountains, ice skating, or even teaching himself to ski—an especially strenuous activity in Wegener's day since lifts to carry a skier up a slope were still unheard of.

After finishing his studies in meteorology, the young

outdoorsman joined the Prussian Aeronautical Observatory as a technical assistant. His principal assignment was to study the upper atmosphere with the help of small instrument-packed kites and balloons. But characteristically, he also made his observations more directly. Together with his elder brother Kurt, he spent fifty-two and a half hours aloft in a free-floating balloon—a record for the time, drifting across many miles of the north German countryside.

In 1906, as word of his courage spread, Wegener was invited to join a Danish expedition to the Greenland ice cap. The trip was a severe test of strength and endurance for all the participants; as no dogs were available, the explorers had to pull their own sleds. The short and stocky young meteorologist endured the extreme discomforts as well as any of the arctic veterans.

Taking up an instructorship in meteorology at the Marburg Physical Institute upon his return, Wegener quickly rose in the esteem of his students. Lecturing in a refreshingly interesting way, he could make clear even the most difficult subjects, such as the complex thermodynamics, or heat flows, of the upper atmosphere. He also endeared himself to his students with his free and friendly attitude toward them; such cordiality was not typical of German university professors of the time. Later, when Wegener became the center of the international debate over continental drift, it was said that his students were so loyal to him that they would gladly have defended his theories with their fists.

Wegener's own enthusiasm for continental drift arose by chance. He knew, of course, about the matching contours of the opposite coasts of Africa and South America. But, looking through a scholarly paper one day,

he learned that there was also a marked similarity between certain ancient bones picked up in Africa and in
South America. This information immediately raised a
question in Wegener's mind. Since these look-alike remains had probably originated from a common ancestor
that could only have lived in a single place, how had its
descendants happened to exist an ocean apart?

While combing the libraries for more information,
Wegener came upon another startling fact. There were
not only strange similarities between extinct creatures on
different continents but also an unlikely across-the-sea
kinship between certain living species. The garden snail
Helix pomatia, for example, was found in both western
Europe and the eastern part of North America, yet the
tiny animal itself could not have made the long journey
across water. Charles Darwin had explained in *On the
Origin of Species* that the snail might have been brought
across the water on the feet of birds, but this was not very
convincing. Pondering the problem, Wegener wondered
whether the continents had not, in fact, once been
connected.

Biologists were certain that they had been connected. But their ideas had nothing to do with continental
drift. Instead, they talked of sunken "land bridges"—
narrow former links between the continents by which
creatures even as poky as snails could have crossed the
seas after generations of effort. One such supposed bridge
was actually given a name: Lemuria, after the lemur, a
little foxlike monkey whose presence in such widely separated areas as India, Africa, and Madagascar (now Mala-
gasy) had long puzzled biologists. Geologists, for their
part, found the whole concept of land bridges slightly
foolish. Except for a possible one-time link between   29

Siberia and Alaska, they pointed out, no one had ever found really positive geological evidence for any such intercontinental connections.

No believer in land bridges himself, Wegener knew that there still had to be an explanation for the presence of similar fossils and animals on different continents. As it happens, an important clue was already available. Long before Wegener began his investigation, geologists had learned that Scandinavia and parts of North America were slowly rising. One sign of this bizarre uplifting was old mooring rings on harbor walls in Scandinavia's ancient Baltic ports. They had over the centuries risen so far above the water that they could no longer be used to tie up ships.

Why should solid earth be lifting? One of the first men to explain the phenomenon was Eduard Suess. It will be recalled that he was mentioned earlier as a drift theorist. Suess knew that there was a fundamental difference in the weight and density of continental land masses, like North America or Scandinavia, and that of the ocean floor which extends beneath them. The first hint of this difference came in 1738. Taking gravity measurements on an expedition to South America, the French mathematician Pierre Bouguer discovered that the gravitational pull on his instruments was significantly less on top of a high Andean peak, Mount Chimborazo, than it was at sea-level sites. The conclusion was that the mountain's granite was lighter and less massive than had been supposed and therefore exerted a smaller gravitational tug. Elaborating on these and other findings about the earth's crust, Suess said that the continents were made of lighter granite-type material called sial (a word coined from *si*lica and *al*uminum, its chief ingredients);

30

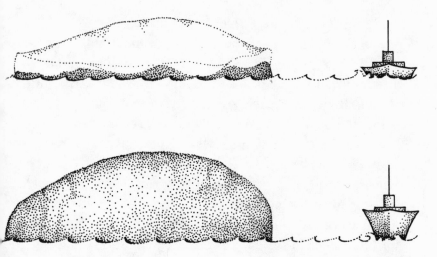

Isostasy at work: (top) *a land mass sinks under a weight of ice, like a ship loaded with cargo;* (bottom) *when the ice melts, the land rises, as the ship does when it is unloaded*

in contrast, the ocean bottom was composed of a heavier, lavalike substance called sima (from *si*lica and *magne*-sium).

Even more important, Suess realized that because of the weight differences, the old principle of buoyancy would apply: the lighter material would, however slowly, eventually rise to the top. Thus, he explained, the continents would float like giant rafts on the heavier ocean rock beneath them. On the other hand, he said, if more weight were suddenly added to these continental rafts, they would sink a little deeper into the water like ships after they have taken on heavy cargo. In geology, this application of the principle of buoyancy is called isostasy (from the Greek words meaning "equal standing").

According to Suess, North America and Scandinavia were giving a textbook display of isostasy at work. During the last Ice Age, he explained, both land masses   31 had been covered by a thick sheet of ice, comparable to the one that still covers much of Greenland and Antarctica. Under its weight, North America and Scandinavia began to sink, like ships loaded to the gunwales with cargo. But, as the Ice Age ended about 11,000 years ago and the sheets of ice melted away, the load on North America and Scandinavia was lightened. As a result, they slowly began to rise. Geologists call such up-and-down movements "isostatic adjustment"; this is simply their way of saying that a land mass is trying to reach the level of flotation required by the principles of isostasy or buoyancy. In the case of Scandinavia, the land is currently lifting at a rate of more than three feet a century, and will continue to rise until the proper isostatic level is achieved.

For Wegener, isostasy offered the solution to the problem of how kindred fossils and animals could have spread over widely separated continents. If continents could bob up and down like corks in water, he reasoned, they should also be able to move sideways, drifting toward or away from each other like giant rafts. Thus, he concluded, it was not the animals that had moved but the continents themselves.

Still, Wegener knew that without actual physical proof his ideas about continental drift would seem no more convincing than any of the earlier ones. Talk of isostasy, fossils, or jigsaw-like similarities in opposite coastlines might persuade laymen but not skeptical scientists. So Wegener began the search for the clinching evidence. Showing his usual dogged persistence, he spent so many hours looking at geological maps, records, and other data that friends warned him that he was jeopardizing his career as a meteorologist.

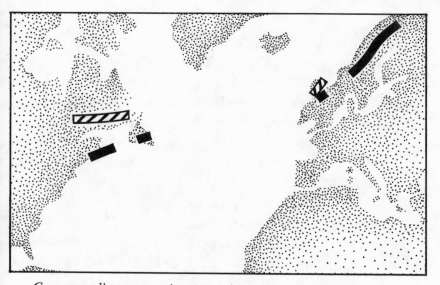

Corresponding mountain ranges in eastern Canada (left) *and in Scotland and Norway*

His patient quest soon produced results. While studying charts showing the world's mountain ranges, he noticed that if he thought of the Old and New World as joined, mountain ranges on opposite continents would form a single, continuous chain. The Sierras, south of Buenos Aires in Argentina, for example, linked up perfectly with the Cape Mountains across the Atlantic Ocean in South Africa. So did mountains in eastern Canada and in Scotland and Norway. Better still, geological dating showed that these separated mountain chains were also formed at approximately the same time in the distant past. The matching mountain chains could be compared to two torn bits of newspaper; not only did the pieces fit together, but the letters also lined up in their proper places.

Now a firm believer in continental drift, Wegener first announced his views in a public lecture in 1912. **33**

Soon, however, he was off on another expedition to Greenland. Showing even more physical endurance than before, he traveled 700 miles by sled across the ice cap and spent a hard winter encamped on a high glacier. He hoped, on his return, to spell out his theory of continental drift in full detail, but the outbreak of World War I in 1914 forced him to put off his work. Though he was deeply distressed by the war, and wondered what effect it would have on international scientific cooperation in the future, he was called up for service in the German army. During the advance into Belgium he was shot in the arm. It was only a minor wound and he soon returned to duty. Two weeks later he suffered a more serious wound in the neck.

The second injury, though painful for Wegener, was something of a blessing for science. During his convalescence, he finally found time to write down his ideas on continental drift. In 1915, at the height of the fighting, they were published in his soon-to-be-famous book *The Origin of Continents and Oceans*.

CHAPTER THREE

Tragedy on the Ice Cap

THE central argument in Wegener's book was that all the continents had once been part of a single, primordial supercontinent—an island that constituted more than a third of the earth's surface. Wegener called this original continent Pangaea, from the Greek words for "whole earth." The rest of the globe, he said, was covered by a vast ocean, which he named Panthalassa, or "universal sea." Far back in time, perhaps as long as 200 million years ago, unknown geological forces inside the earth began to disintegrate Pangaea. Slowly, its fragments started to drift apart—the Americas heading west, Asia north, Australia east, and Antarctica south. Only Europe and Africa remained relatively stationary.

35

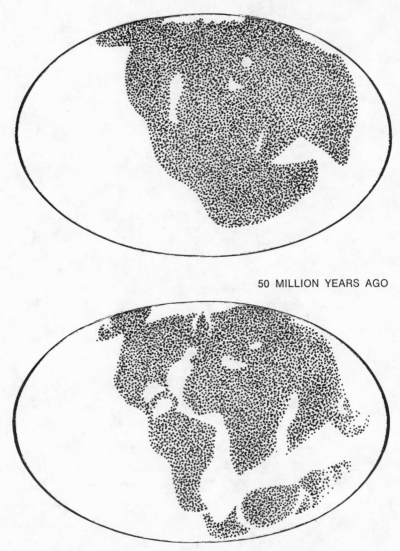

300 MILLION YEARS AGO

50 MILLION YEARS AGO

Wegener's reconstruction of the break-up of Pangaea

Revolutionary in its scope, Wegener's theory answered many troublesome geological questions:

Q. Why did the continents seem to fit together?
A. They were all once part of a single supercontinent.
Q. Why were related plants and animals found widely scattered across the globe?
A. They had been dispersed when the continents drifted apart.
Q. How did tropical fossils ever find their way to extremely cold climes?
A. They had been carried by the continents as they moved from one climatic zone to another—a prime example being India's drift from a location south of the equator to one north of it.

These were only some of the more obvious problems solved by continental drift. Wegener's theory also touched on one that, at first glance, seemed completely unrelated to drift: the origin of the earth's great mountain ranges. In the nineteenth century, geologists had assumed that the earth was once a hot, fiery ball. As it cooled off, they said, it hardened and contracted, leaving its outer layers wrinkled or folded over, like the skin of a shriveled apple. It was such folding of the earth's crust that was assumed to have produced mountains. But in Wegener's time, this mountain-building theory was being questioned seriously. By then, geologists had made careful studies of existing mountain ranges and found that they showed far too much folding. It would have required considerably more cooling and contraction than the earth could possibly have undergone. Furthermore, the recent discovery of radioactive elements like uranium had cast a shadow over the whole concept of a cooling

37

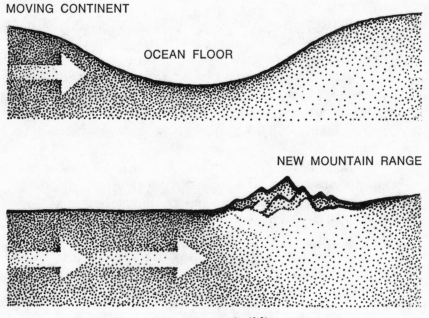

MOVING CONTINENT

OCEAN FLOOR

NEW MOUNTAIN RANGE

Wegener's theory of mountain building

earth. Because such materials give off heat, some scientists suspected that the earth might in fact be heating up and even expanding. How, then, geologists asked themselves, were mountains formed?

Wegener's theory again offered an answer. It was not contraction or expansion of the earth that had created mountains, Wegener said. Instead, it was the movement of the continents. Pushing through the ocean floor, he explained, the continents encountered increasing resistance from it, like a ship struggling through a thick ice field. After a time, so much resistance built up that the front edge of the continent crumpled, folded back, and was actually thrust up. The result of this pile-up of crust at the continent's edge was the birth of a mountain chain. As an example of such mountain-building, Wegener pointed to the Andes and the Rocky Moun-

38

tains. Forming a lengthy, almost continuous chain along the western coast of the Americas, these peaks were presumably created as the continents moved westward against the increasing resistance from the floor of the Pacific Ocean.

Elegant as it was, Wegener's theory did not win universal acclaim. Although a few scientists acknowledged its merit, most geologists found it difficult to accept. In part their skepticism was due to scholarly snobbery: they felt that since Wegener was not a geologist he was totally unqualified to speak on geological matters. But they also had a number of more serious objections to his theory. In the first place, they wanted to know how continents made of lighter, softer rock could penetrate the harder, more dense rock of the ocean floor. Wegener had explained that the tougher ocean rock gradually yields, or gives way under the enormous pressure against it. But his critics replied that this was like arguing that a saw made of soft iron could cut through tough steel.

In addition, they questioned Wegener's dating of Pangaea's breakup. Since the age of the earth was known to be several billion years, they asked why the breakup had occurred only 200 million years ago rather than earlier in the earth's long history. Some scientists even disputed Wegener's analysis of the matching shapes of the continents. They said that most of the supposed similarities in the coastlines were at best highly exaggerated; in cases where there was an undeniable match, as between Africa and South America, they insisted that it was no more than a coincidence. In contrast, other critics accepted the fitting together of the continents but ironically used it as an argument *against* Wegener's theory. If an event violent enough to break up Pangaea had actually occurred, they contended, it would have crushed so much material that the remaining pieces would be no more likely to match than the bits of a smashed cracker.

Yet the strongest objections to Wegener's theory came from geophysicists, scientists who specialize in studying the forces acting on the earth. They wanted to know what power could possibly propel continents through the hard ocean floor. Wegener had suggested that the earth's own rotation provided some of the power. As the earth spins on its axis, he explained, the motion tends to make its outer layers fly off. This same centrifugal force is experienced by passengers in a car or on a train when they slide in their seats as it rounds a curve. Clearly, the earth's rate of spin is not fast enough to cause pieces of the crust to fly off into space. But, said Wegener, it could push material—including continents —away from the poles and toward the equator. As proof of this pushing, which he called by its German name *Polflucht,* or flight from the poles, he pointed to the earth's slight equatorial bulge. *Polflucht,* however, was not his only power source for continental drift. Trying to account for the westerly drift of North and South America, Wegener offered a simpler explanation: that movement, he said, was the result of the gravitational pull of the moon and the sun—the same pulling that causes the ebb and flow of the ocean tides.

Geophysicists admitted that Wegener's *Polflucht* and tidal forces were real enough, but their calculations showed that he had vastly exaggerated their strength. Taken individually or together, these forces exerted no more than a fraction of the power that Wegener attrib-

40

uted to them. For the northerly drift of India, they offered no explanation. In 1928, at a historic scientific meeting on continental drift, one geophysicist argued that it would have taken a million times more power than *Polflucht* could deliver to have nudged the continents a few miles, to say nothing of moving them across the globe, over the course of geological time.

Wegener's response to these devastating attacks was typical of the man. He good-naturedly conceded that some of the criticism was justified—for example, the inadequacy of his mechanisms for moving the continents. But on the overall question of whether continental drift had actually occurred, he stubbornly refused to budge. Instead, he intensified his search to prove the truth of his theory.

In some ways, this single-minded determination did him more harm than good. For one thing, it made him prone to overstate his case, seeing evidence in favor of drift where none existed. By comparing star sightings that had been taken over the years in the arctic, he calculated that Greenland was drifting as much as forty yards annually; at that improbable clip the continents would be circling the earth once every two million years —or a hundred times since Pangaea's breakup!

He was also hurt more personally. For years, he was denied a regular professorship in the German universities because he was thought to be dabbling too much in an area outside his own professional discipline, meteorology. When he was finally appointed to a professorial chair at the age of forty-four, it was not at a German university but at Graz University in neighboring Austria.

Unfortunately, there was little time left for Wegener to defend himself against his critics. In the late 1920s   41 while the debate over his theory swirled around him, the veteran scientist-explorer organized another expedition to Greenland; one of his objectives was to measure the thickness of the ice cap. He was beset with problems from the start. Even before the expedition's departure in the summer of 1930, it was almost canceled for lack of money. Then, after it finally got under way, Wegener's ship had to wait six weeks off Greenland until a suitable landing spot was found in the ice-packed coast. As the expedition's leader, Wegener was occupied with a never-ending round of chores. He had to bargain with the Eskimos for sled dogs, hunt for additional supplies, and make several risky trips from his coastal headquarters to the edge of the ice cap, checking the routes to the expedition's isolated outposts.

His spirits never sagged. Late in autumn, in spite of a raging storm, Wegener set out by sled to bring food to the expedition's mid-ice meteorological station. The trip took nearly six weeks; it snowed almost continuously and temperatures fell as low as minus 65° Fahrenheit. But Wegener showed no sign of any physical weakening. After he had assured himself that the scientists at the station could keep it open through the winter, thereby gathering valuable weather data, Wegener and an Eskimo companion started back for their base camp on the coast.

Although Wegener thought the return trip would be much easier, he and his companion soon found themselves trapped in another severe blizzard. Day after day, the two men battled to make headway against the lashing winds and snow, but it was a losing fight. Finally, even Wegener's robust body was weakened by the unrelenting cold, and he could go on no more. Miles away

42

from camp, Wegener died on the ice cap and was buried at the lonely site by his Eskimo friend. Saddened members of Wegener's expedition did not recover his body until the following spring when the thaw melted away some of the snow.

After Wegener's tragic death at the height of his career, a few scientists continued to argue in favor of continental drift. The South African geologist Alexander L. du Toit and the Scottish geologist Arthur Holmes, in particular, tried to keep the theory alive. Rock formations in Africa, South America, and Antarctica, they insisted, showed too many similarities to be dismissed as mere coincidences. They even tried to save the major elements of Wegener's theory by devising slightly different (and less objectionable) ways in which the continents might have been moved. But these efforts were futile.

With the death of its major spokesman, continental drift soon was all but forgotten, as was Wegener himself. If it was mentioned at all, it was usually with contempt. Geology professors occasionally held it up to their students as a classic example of scientific blundering: using doubtful observations (that the continents have similar coastlines) to support an unfounded conclusion (that the continents were once a single land mass). Indeed, by the late 1940s, continental drift was so discredited, especially in United States scientific circles, that any geologist who argued for it risked the open ridicule of his colleagues and, like Wegener, his chances of obtaining a university teaching job.

Yet, to paraphrase Mark Twain's famous comment on the false accounts of his own passing, the reports of the death of continental drift were slightly premature.　43

In the early 1950s, more than twenty years after Wegener's fatal expedition to Greenland, the old ideas were suddenly revived. The reason for the renewed interest was that surprising new evidence in favor of continental drift had come from a totally unexpected quarter: the study of magnetism in ancient rocks.

Alfred Wegener in 1919

P. M. S. Blackett

Maurice Ewing in front
of the research vessel
Vema

Alfred Wegener in 1919

P. M. S. Blackett

Maurice Ewing in front
of the research vessel
Vema

Bruce C. Heezen

Harry H. Hess

Lynn R. Sykes studying seismographs

Robert S. Dietz (left) *and John C. Holden*

CHAPTER FOUR
The Secret of the Ancient Compasses

PERHAPS the momentous discovery was made long ago
by some shaggy-haired Neanderthal man. Upon ventur-
ing out of his cave one day, he picked up two odd stones
and started playing with them. To his astonishment, he
found that when he held his new toys in one position
they pulled together. In another position, they pushed
apart. No one can tell for sure if this was how man made
his initial acquaintance with magnetism, but it is known
that the ancient Chinese were probably the first to put
such natural magnetic rocks to any practical use. By
suspending such a rock freely on a string, they found
it always pointed north. In this way, the compass was
born.   49

Some scholars now think that the ancient Maya of Central America may also have known how to make such primitive compasses. The word magnetism itself comes from the name of a district in ancient Greece, Magnesia, where many magnetic rocks were found. In Europe, iron-rich magnetic rocks were called lodestones (after the Old English word for course or way). Bad weather or good, lodestone compasses would faithfully show lost sailors the right way. Why these compasses always pointed north was not understood, although some medieval scholars guessed (incorrectly) that Polaris, the North Star, was exerting an attractive force on them. It was not until 1600 that William Gilbert, physician to England's Queen Elizabeth I, finally solved the mystery.

A shrewd, self-taught experimenter, Gilbert brought magnets and compasses of various sizes and shapes into his laboratory and then began the first thorough scientific investigation of magnetism. Out of these pioneering experiments, he drew certain fundamental conclusions. Every magnet, he said, has two poles: north and south. Furthermore, like poles (for example, two norths) repel or push each other apart, whereas unlike poles (a north and a south) attract or pull together. These attractive and repulsive forces are not equal in magnets of different sizes, he found. If he happened to pass the needle of a tiny compass along the surface of a much larger lodestone, only the compass needle moved; the heavier lodestone remained still.

These fundamental ideas about magnetism may sound simple today, but they finally explained why a compass points north. The earth, said Gilbert, is itself an immense magnet with north and south poles. Like those of the lodestone in his laboratory, the earth's poles

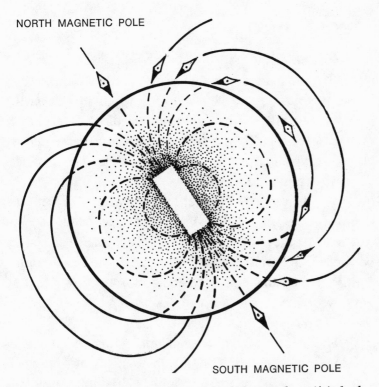

SOUTH MAGNETIC POLE

Lines of magnetic force sweep around the earth as if it had an ordinary bar magnet in its interior

attract or repel the poles of smaller compasses, yet show no movement themselves.

Gilbert did not know why the earth acted like a magnet, although for a long time people speculated (again incorrectly) that there was a rich vein of lodestone near the North Pole. Only recently have scientists become reasonably certain that the earth owes its magnetism to internal electrical currents in its core, apparently created by effects of the earth's rotation on this molten metal. Gilbert, however, was sure of one thing: the earth's invisible magnetic field was similar to that of an ordinary bar magnet, or what scientists call a dipole—the lines of force spreading out from its north pole and then coming together at its south pole.

How could he tell this? Gilbert shaped a lodestone into a ball and imagined it to be the earth. Then he took a compass needle and held it over various places on the lodestone's surface. Invariably, the needle pointed in the same direction as the lines of magnetic force created by his miniature "earth." At the North Pole, for instance, the needle pointed straight down. At the equator, on the other hand, it rested parallel to the surface. In between these two positions, the needle's dip varied; the closer it was held to the North Pole, the greater the needle's dip became. Gilbert knew that he had succeeded in duplicating the earth's magnetic field, because sailors had already observed such dipping of their compass needles as they sailed over the globe.

The sailors of course had no inkling of what was causing this magnetic dip, or inclination, as scientists call it, but such magnetic readings do in fact provide useful navigational information. The dip can tell a sailor his latitude, or distance from the poles: if the compass needle showed an inclination of 90°—that is, pointed straight down—it would mean that he was right over the North Pole. If the needle did not dip, or showed an inclination of 0°, it would indicate that he was at the equator. If the inclination read 45°, or halfway between the other readings, it would mean that the sailor was somewhere between the North Pole and the equator. For the sake of simplicity, it is being assumed that the earth has only one North Pole; actually, it has both a geographic and a magnetic north pole. They are hundreds of miles apart. To navigate accurately, a sailor has to take the separation into account, since his compass always points to magnetic rather than geographic north.

52 Inclination can also be used in the southern hemi-

sphere. There it is the south-seeking end of the compass needle, rather than the north-seeking one, that does the dipping.

As it happens, magnetic inclination is not only helpful to sailors at sea but is also an extremely useful tool for paleomagnetists (from the Greek *palaeo,* or old), scientists who study the magnetism of ancient rocks. Such rocks are, in effect, fossilized compasses. They are excellent indicators of the direction of the earth's magnetic field that prevailed at the time and place of their formation, because the earth's magnetic field is permanently locked into the rock's atoms as it cools off and hardens. Even if the rock is later moved or the earth's magnetic field changes, the rock will still retain this original magnetic imprint. Some of the best imprinting occurs in lava because it is rich in iron. Scientists have known about such faint fossil magnetism for more than a century. But since it was barely detectable, they did not pay much attention to it until the recent development of an extraordinarily sensitive measuring instrument called an astatic magnetometer.

Invented in the early 1950s by a British Nobel laureate in physics, P. M. S. Blackett, the magnetometer could detect the direction of extremely weak magnetic fields. It would respond even to a field that was only one ten-millionth as intense as the earth's. Paleomagnetists were overjoyed with the device. Long tantalized by rock magnetism but unable to analyze it easily, they now had in their hands the key to unlocking the secrets of the ancient rock compasses.

British paleomagnetists, notably S. Keith Runcorn, were the pioneers in this field. Their first important finding came while they were studying layers of rock in the

53

English countryside formed some 200 million years ago, in the Triassic period. According to their measurements of the rocks, England's average inclination at that time was about 30°. At present, England's inclination is 65°. How could this great difference be explained? Their answer would have delighted Alfred Wegener. England, they said, must have been located much farther south during the Triassic period because 30° indicates a position closer to the equator than 65°.

With proper scientific caution, they decided to confirm their astounding data with another test. This time they analyzed rocks from India's sprawling Deccan Plateau. Covered with strata of many different ages up to 150 million years old, the region is like a geological history book with each layer providing a different chapter in the story.

This history book told them that in rocks of the Jurassic period, 150 million years ago, India's average magnetic inclination was 64° south; that is, a compass needle at that time and place would have dipped toward the South Pole. In rocks of the Cretaceous period, 100 million years ago, however, India's inclination had decreased slightly to 60°. Even more interesting, by the early Tertiary period, 50 million years ago, it had dwindled to a mere 26°. Finally, in the middle Tertiary period, about 25 million years later, India's inclination did a complete flipflop and now was 17° north—that is, the compass needle would have dipped toward the North Pole.

For the paleomagnetists, this evidence was even more spectacular than their original findings in England. What it seemed to say was that India had once been located below the equator in the southern hemisphere.

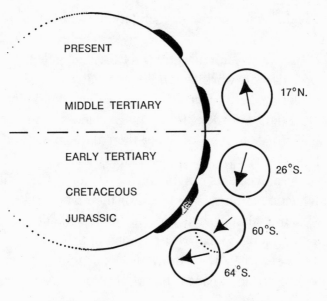

India's position in various geologic periods

Then, judging from the gradual decline in the southerly inclination readings, it edged slowly north. Eventually it apparently crossed the equator (causing the reversal to a northerly inclination) and ultimately arrived at its present position. Could these readings mean that one of the details of Wegener's forgotten theory was really true? Had India gradually drifted north?

Not likely, said other scientists. They refused to accept the notion that India or England had drifted. Instead, they pointed out that the varying readings could have been caused just as easily by changes in the position of the North Pole itself over the course of the earth's long history. Such polar wandering, they explained, would also have altered the direction of the earth's magnetic field in any one place; thus rocks formed at different times in India or England could well have been imprinted

55

with different magnetic inclinations without moving at all.

In fact, another group of British paleomagnetists were already working on the premise that such polar wandering had occurred. They had made similar magnetic studies of rocks found in England and in continental Europe. Instead of using their data to show drifting of land masses, they painstakingly calculated how the pole's movements might have produced the different magnetic readings. The result of their work was a sweeping curve that traced out the pole's wandering across the top of the globe over millions of years.

Yet, when they drew a second curve from rocks picked up in various parts of North America, they

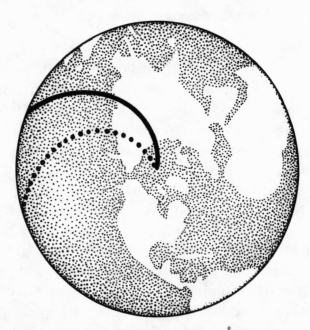

Movement of the north magnetic pole as indicated by magnetism in North American rocks (solid line) *and in European rocks* (dotted line)

also encountered a surprise. Although the two polar paths took roughly the same shape and had a common point of origin, the curves gradually veered away from each other. Why should there have been such a wide separation in the curves drawn from the European and American rocks? If only the pole had moved, and not the continents, then the magnetic imprint left by such wandering should be the same in rocks of the same age all over the globe.

The paleomagnetists did not have to search long for an answer. They found that when they imagined the Atlantic Ocean as closed up and Europe and North America as joined, the two separate curves now overlapped almost perfectly. With that imaginative move, they also closed the gap between themselves and the other group of rock magnetists. For while the curves were good evidence for polar wandering, they also provided a strong hint of continental drift.

Even so, not everyone was convinced. For one thing, the theory of continental drift had been discredited for so long that it could not be revived overnight. For another, paleomagnetism was still a relatively new science; the paleomagnetists themselves admitted that their readings were subject to error. Some scientists suggested that the differences in magnetic inclination might have alternate explanations—for example, the existence of multiple poles at certain times in the earth's past.

However, the work of the magnetic fossil hunters could not be dismissed. On the contrary, by the late 1950s, they had accumulated so much data that Wegener's theory was again becoming a topic of serious scientific discussion. Though far from proved, continental drift had now been rescued from the scientific scrap heap. 57

CHAPTER FIVE

Probing the Ocean Depths

THE shrewdest detectives are unable to solve a mystery if they ignore the most important clue. In trying to unravel the mysteries of the earth's surface, geologists have also overlooked significant signs. By concentrating only on the ground directly underfoot, the biggest clue of all escaped them: the three-fourths of the earth submerged under the seas.

Their neglect was partly because of indifference. While there had long been talk of sunken continents and of lost civilizations like the legendary Atlantis, many scientists considered the ocean floor as a place devoid of interest. And they had good reason for their attitude. In the few areas where it had been examined it seemed barren and featureless, covered with piles of muddy sediment. However, if the early geologists had wanted to take a close look at the ocean floor, they probably could not have done a thorough job. Such undersea exploration required far better ships and tools than they had at their disposal.

As so often happens in the search for truth, the first great advances in marine geology came as a by-product of ventures far afield from science. In 1854, the U.S. Navy began taking measurements of the Atlantic Ocean's depth in preparation for laying a telegraphic cable between Europe and America. During these soundings, the Navy men learned that the middle of the Atlantic, instead of being the ocean's deepest part, was unexpectedly shallow. This surprising mid-ocean rise became known as Telegraph Plateau—a name that stuck until a later, more momentous discovery occurred in this part of the Atlantic.

The real beginning of oceanography came when Her Majesty's Ship *Challenger,* a full-rigged British Navy corvette of 2,306 tons with auxiliary power, set sail on a three-year world cruise (1872–1876). One of the *Challenger*'s objectives was to gather information on the ocean floor to help lay telegraph cables. But the expedition accomplished far more than that; in the course of the long and difficult voyage (during which two men died and others deserted) the *Challenger's* scientists collected so much information about the oceans that it took twenty years to catalogue and filled fifty volumes. Among other things, the *Challenger* found that there are many "hills" under the sea and that it sometimes took nearly four miles of measuring line to reach the ocean bottom. As they studied the *Challenger*'s reports, scientists began to realize that the ocean floor is much more complicated—and interesting—than they had thought.

Undersea exploration took its next great plunge forward after World War I, thanks mainly to the development of a new instrument that enabled scientists to "see" the ocean floor for the first time. It was the echo sounder, a device for measuring depth with sound waves. Before the invention of the sounder, the only way to determine depth was to do what the *Challenger*'s crew had done a half century earlier: halt the ship in mid-ocean, heave a measured lead-weighted line over the side and play it out until it touched bottom—if possible without breaking. The new sounder automated the job: by bouncing the waves off the bottom and clocking the time it takes them to return, the gadget could quickly calculate the depth (since the speed of sound through water is known).

The first vessel to make serious use of the echo sounder was the German oceanographic research ship *Meteor*. The Germans, too, were not interested in science alone. Hoping to find an easy way to pay off their debt for World War I, the Germans thought the *Meteor* might be able to uncover rich deposits of gold in sea water. The idea was not entirely preposterous; each cubic mile of ocean contains millions of dollars worth of gold, although the cost of extracting it would run far higher. The *Meteor*, alas, did not find nearly enough watery ore to end Germany's financial headaches, but it did help to produce a big scientific payoff. Using the new echo sounders, the *Meteor* and other pioneering oceanographic vessels made extensive sonic profiles, or silhouette-like pictures of the ocean floor, during their far-flung voyages in the 1920s and 1930s. These profiles gave geologists a truer image of the ocean floor. It was an imposing landscape—with 61 plains as broad as those of Kansas, mountains as high as the Rockies, and trenches deeper than the Grand Canyon.

The grandeur of the underwater world was only one surprise. Geologists have long been convinced that the oceans are the earth's most ancient feature, dating back 2 billion years or more. Accordingly, the ocean floor should be thickly carpeted with sediment, the muddy debris of dead marine organisms and wastes washed off the continental shelf. So rapidly does such sediment build up that even after 100 million years, a relatively brief interlude in the history of the oceans, it should be at least a mile thick.

To put their calculations to the test, scientists perfected a technique called "seismic shooting"—exploding small satchels of dynamite behind their ships, thereby creating tiny manmade earthquakes at sea. As the seismic waves travel through the ocean floor, their speed varies in material of different density; the waves slow down in the softer, looser sediments and pick up speed in the harder, denser rock beneath them. Thus, by clocking the waves, the scientists could get an idea of the thickness of the different layers on the ocean floor. However, when they began tallying up their results, they got quite a jolt. Instead of miles of sediment, they found an average thickness of a mere 1,500 feet. Did this mean that the oceans were much younger than anyone had supposed? Or, the geologists asked themselves, was some kind of natural vacuum cleaner sweeping the ocean floor clean?

Before they could even begin to answer these questions, many more popped up. By towing highly sensitive magnetometers behind their ships, oceanographers found that the magnetic patterns locked into the sediment alternated with bewildering regularity. In some areas, these

fossilized compasses pointed north; in others, they pointed south. They had reversed! Gravitational measurements were no less confusing. It was known that the earth's crust was being pulled down into the deep trenches located at the edge of some continents. Yet the measurements taken in the trenches showed that the tug of gravity was too small to account for the pulling. A hitherto unknown force was at work in the trenches. Then there were curious hot spots in mountainous regions like the middle of the Atlantic. Why was heat seeping out of the earth at so prodigious a rate in these areas? The more oceanographers explored the ocean floor, the more complicated it became.

Late in the 1950s, oceanography reached a peak of activity. Under the sponsorship of the U.S. Navy, oceanographic vessels had been busily crisscrossing the seas in an intensive effort to chart the floor of all the world's oceans. The reason for such undersea cartography —which was also being pursued by the Russians—was the need to provide maps for the growing fleets of deep-diving nuclear submarines. By 1958 so much data had been compiled that oceanographers Bruce C. Heezen and Marie Tharp of Columbia University's Lamont (now Lamont-Doherty) Geological Observatory could come to a startling conclusion. The submerged mountains and undersea ridges, they announced, form a continuous 47,000-mile-long worldwide chain which circles the globe like the stitching on a baseball. Though buried under water, this great "mid-ocean ridge" was easily the dominant feature on the face of the earth, extending over an area larger than that of the Himalayas, Rockies, and Andes combined. "Imagine millions of square miles of tangled jumble," said Maurice Ewing, Lamont-Doherty's

63

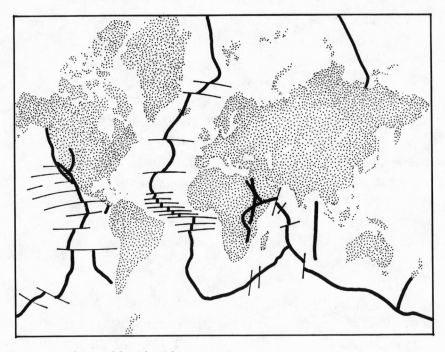

The world-wide ridge system

director and a long time explorer of this forbidding under-sea world. "Massive peaks, saw-toothed ridges, earth-quake-fractured cliffs, valleys, lava formations of every conceivable shape—that is the Mid-Ocean Ridge."

For all its magnificence, the ridge system posed additional problems. Along much of its length, scientists learned, the ridge is cleaved right down the middle by a sharp break, or rift—which is the center of intense heat flows. Moreover, the ridge is the site of earthquakes and periodic volcanic eruptions, such as the one that created the new island of Surtsey, off the coast of Iceland, in 1963. Was the system a giant crack in the earth? If so, it was a puzzling crack indeed. As the ridge runs crazily around the world it is crossed at right angles by dozens

64

of deep faults, akin to those caused by earthquakes. Along these faults odd, straining movements seem to be taking place, as if the ocean floor were moving. Finally, although the ridge is steep and jagged in the Atlantic, it takes on a much gentler look in the Pacific and Indian oceans, where it sends off smaller branches that slip under the continents themselves.

Scientists were not at all sure what this meant, but they were beginning to suggest some answers. At the very least, the ocean floor was far more active than had been supposed. It was also apparently much younger, especially at the ridges. Furthermore, the indications of volcanic activity hinted that the ridge itself might be adding material to the ocean floor. In contrast, the trenches off the continents suggested an entirely different process; they seemed to be swallowing up oceanic material. This then might account for the missing sediment and perhaps for the strange gravitational readings.

Beyond this, there was the intriguing position of the ridge itself. In the Atlantic Ocean, it follows a winding route midway between the opposing continents, not just Africa and South America but also Europe and North America. During Wegener's day, many scientists had dismissed the suggestion of dovetailing of continents as a geographical accident. Now, with this new evidence of still another line running exactly in the middle of two other parallel lines, the matching coastlines could no longer be so easily brushed off. Skeptics were now beginning to ask if Wegener's ideas did not contain some shred of truth after all.

CHAPTER SIX

Conveyor Belts in the Sea

WHILE he was the commander of a U.S. Navy transport in the Pacific during World War II, Princeton University geologist Harry H. Hess did not forget his peacetime interest. One day, while scannning readings from his ship's echo sounder, he noticed the profile of an unusual seamount, or submarine volcano. Its top was completely flat, almost as if it had been sawed off. Hess soon located a total of twenty such flat-topped peaks, which he dubbed *guyots* (pronounced "ghee-oh") in honor of a distinguished nineteenth-century Princeton geologist, Arnold H. Guyot. Other scientists eventually found hundreds more guyots throughout the Pacific, and speculated

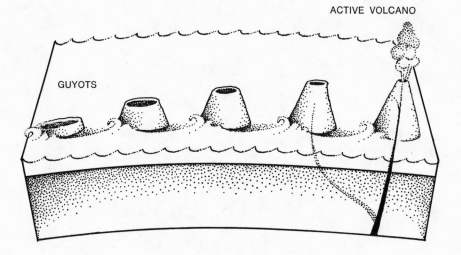

ACTIVE VOLCANO

GUYOTS

Guyots are reduced in size by wave action as they move away from the volcanically active area in which they originated

that their cones had gradually been worn away by the pounding of the waves.

Still more intriguing things were learned about these odd-shaped undersea mountains. Scientists realized that the farther a guyot was from volcanically active areas of the ocean, the older and flatter it seemed to be. This observation raised an interesting problem: had the guyots actually wandered across the sea floor away from their places of birth?

Like the majority of his fellow geologists, Hess was no more inclined to believe in drifting mountains than in moving continents. But as the chaotic picture of the ocean floor became clearer, he began to re-examine his doubts. Something had to be churning away inside the earth to produce such upheavals on its surface. Could these same powerful engines be moving the volcanoes— possibly the continents?

For a long time, Hess wrestled with his deeply ingrained doubts about the Wegenerian theories. But by the early 1960s, he gave his answer. Reviving some long-forgotten ideas, changing them a bit, and adding a few insights of his own, he proposed a scheme so at odds with accepted geological fact that he cautiously gave it the name "geopoetry."

Hess's geopoetry began with the prevailing theory of the origin of the earth itself. Although no scientific question has been the subject of more avid debate, many scientists agree on one possible version of that unwitnessed drama: Five billion years ago, when the sun was still in its infancy, a huge, dense cloud of cosmic dust gathered around the young star under the pull of its powerful gravitational field. As the dust particles swept around the sun in elliptical orbits, many of them came close to one another, collided, and in some cases stuck together. These enlarged fragments were hit by still more particles, and gradually the clumps increased in size. At last, after countless such collisions, the moons and planets of the early solar system were born.

The newborn earth and its young celestial neighbors were originally relatively cold and brittle. But since the earth's stockpile of radioactive elements gave off heat, just as does similar material in a modern nuclear power plant, the primordial planet rapidly warmed up. By the time it was a youthful billion years old some melting had started inside the earth. Then there began a geological process called differentiation: heavier elements, like iron and nickel, sank swiftly in a molten mass to form the earth's 2,100-mile-thick core, which itself consists of a solid inner part surrounded by an outer layer of molten material. Lighter elements, such as silica, rose to create

69

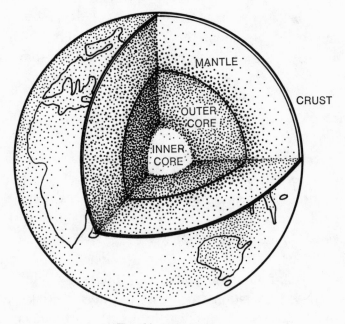

Earth's interior

the multiple layers of the 1,800-mile-thick mantle and the earth's 25-mile-thick eggshell of a crust. On top of all this, a slag of granite congealed, like a patch of cream afloat in a bowl of milk. This formed the original continental material.

This "cold" version of the earth's birth, which has largely replaced the theory that it was born out of hot gases, holds that the inside of the planet is still being kept warm by radioactivity. Surprisingly, the center of this nuclear heat production is not the core but an upper region of the mantle, apparently rich in radioactive elements like thorium and radium.

Called the asthenosphere (from the Greek word *asthenos*, weak), the layer is really not weak at all by ordinary standards. In fact, if man ever managed to drill

70

far enough into the earth and bring up a sample, the asthenosphere's rock would probably seem hard. Still, over long periods of time, the asthenosphere gives the impression of weakness. As a result of the radioactive heating, the rock becomes slightly plastic, though not melted. It rises slowly, moves to the side, and then sinks back down. Such movements, similar to those of oatmeal simmering in a pot, are called convection currents.

That rock can flow without melting sounds strange. In fact, many materials, including the toughest steels, exhibit such a tendency under high enough temperatures and pressures. The blades of jet engines, for example, slowly deform with continued use, even though the steel has not really softened. Perhaps the most familiar example of such "creep," as engineers call it, is the behavior of the toy substance called Silly Putty. If Silly Putty is molded into a ball, it will bounce like the hardest rubber. If it is left standing overnight, it will collapse and flatten out into a shapeless glob.

Hess knew that creep could occur in the earth. He also recalled that as far back as the 1930s, Wegener's old champion, Arthur Holmes, had suggested such slow, sluggish movements in the mantle to account for continental drift. The drifting of the continents, Holmes had insisted, is too obvious to deny. The pioneering Dutch geophysicist Felix A. Vening-Meinisz, who took gravity readings in the ocean trenches from deep-diving submarines in the 1930s, also believed in the existence of such convection currents. But since no one could peer into the mantle and convincing evidence on the earth's surface was lacking, scientists rejected the idea.

Then along came the discovery of the mid-ocean ridge system in the late 1950s. These hot, volcanically active crevices in the earth's crust demanded an explanation. Had the earth expanded and cracked open like a boiling egg? Bruce Heezen, the co-discoverer of the ridge system, offered a tantalizing theory. The earth's supply of internal heat, he explained, was inadequate to cause the planet to expand. But, he added, the earth might have grown larger if there had been a decrease in the force of gravity, a possibility long considered by physicists. Since gravity is the "glue" that holds the earth's material together, any weakening of this basic binding force would cause the globe to bulge out, like the stomach of a fat man who loosens his belt.

Nonetheless, most scientists found Heezen's expansion theory unconvincing; gravity, like the speed of light, seems to remain constant over long periods of time, and no one has ever been able to prove that it is changing. So, if the earth did not expand, how were the ridges formed?

In his now classic paper "Evolution of the Ocean Basins," Hess gave his geopoetic answer. The mid-ocean ridges, he agreed, are indeed a kind of crack in the earth's crust, but they are not caused by expansion. Instead, he explained, they owe their origin to convection currents, the old idea of Holmes and Vening-Meinisz. As the rock heats up in the asthenosphere, Hess explained, it inches slowly upward, just as does oatmeal in a simmering pot. Finally, after millions of years, the rock reaches the topmost layer of the mantle, called the lithosphere (from the Greek *lithos,* rock). At this point, because of the reduced temperature and pressure, the original rock—a greenish mineral called olivine—undergoes a physical transformation and changes into a veined rock named serpentine. Some of the serpentine attaches itself to the lithosphere; the other part continues upward, actually pressing through

72

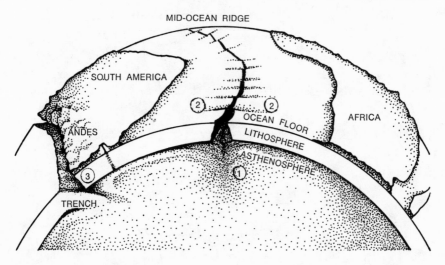

Mechanism of sea-floor spreading: hot material rises at a mid-ocean ridge (1); after solidifying, the new rock moves away from the ridge in opposite directions (2); material eventually presses back into the earth at a deep-sea trench (3)

the seventy miles of lithosphere until it approaches the ocean floor. Then in another change, the serpentine melts, breaks the earth's crust at the mid-ocean ridges, and pours out as molten lava.

The lava does not stay melted for long. As it cools it joins up with the older material at each side of the ridge and hardens into the familiar basaltic rock of the ocean floor. Under pressure from material rising up underneath, the rock continues to press outward in opposite directions. So, too, does the serpentine that has attached itself to the underlying lithosphere. In fact, both layers—ocean floor and lithosphere—act as a single plate that is continually moving away from the mid-ocean ridge.

If a continent happens to be embedded in this plate, 73

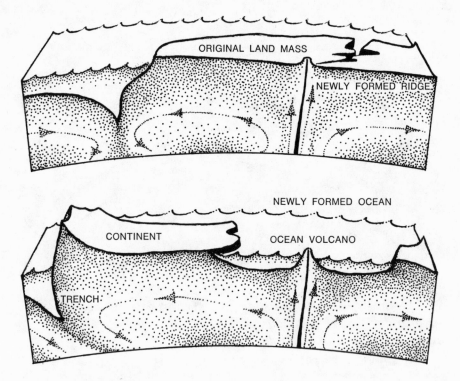

Birth of a continent: (top) *material welling up at a new ridge breaks the original land mass apart;* (bottom) *the newly formed continent gradually moves away from the volcanically active ridge*

it will simply be carried along by it, just as if it were riding on a giant conveyor belt. Since the earth's surface remains roughly the same size, however, any material added to the sea floor at one end of the belt must be matched by a subtraction at the other. Hess suggested that such destruction of the old sea floor and the lithosphere took place at the deep-sea trenches, which, as has been noted, are usually found at the edge of continents or along such volcanic island chains as Japan and the Philippines. There, millions of years after its formation at the mid-ocean ridges, the cooling plate presses back into the earth, presumably to be broken up, remelted, and re-absorbed into the mantle for a renewal of this age-old earthly cycle.

The elegance of Hess's presentation impressed many scientists. Not only did it explain the origin of the mid-ocean ridges and other puzzling features of the ocean; it also provided the long-sought mechanism for continental drift. No longer was it necessary to argue that the lighter continents plowed through the earth's denser, heavier crust like Wegener's ships in a frozen sea. Instead, Hess suggested a far more likely kind of passage: the continents were rafted along, like giant ice floes in a moving current.

Still, Hess's theory of sea-floor spreading—as it was called by the American geologist Robert S. Dietz, who advocated the same general idea in a slightly different form—did not resolve the old argument over continental drift. Beautiful as his geopoetry was, without actual proof it was a long way from geological fact.

CHAPTER SEVEN

Testing the Geopoetry

EAGER as they may be to explain nature's puzzles, scientists are often very cautious people. Privately they may like a new theory, but publicly they may express doubt about it. Einstein's general theory of relativity, for instance, is still the subject of debate among physicists, although it is unquestionably one of the great achievements of twentieth-century science. Not that physicists or other scientists enjoy being obstinate; their desire is to determine the truth of any fresh idea. One of their favorite ways of doing this is to pit the idea against actual observations. In particular, they like to look for some natural oddity that would be hard to explain except by the new theory.

In 1963, a graduate student in geology at Cambridge University devised just such a test for the Hess-Dietz theory of sea-floor spreading. Frederick J. Vine, then only twenty-four years old, had been fascinated ever since he was a teen-age schoolboy in England by the concept of continental drift. He realized, though, that more than the jigsaw fit of opposing coastlines was needed to make "drifters" of his fellow geologists. Now, as earth scientists debated the pros and cons of Hess's geopoetry, the young man was struck by one of those rare flashes of insight that turn out to be important milestones in the march of science.

Vine realized that there had been periodic reversals in the earth's magnetic field: that is, the North Pole became the South Pole and vice versa. Such switches are still a puzzle to scientists. But they do know from the magnetic record in ancient land rocks that the poles have reversed a number of times in the past 3.5 million years. More important, an even longer record of magnetic reversals has been left on the ocean floor. By towing magnetometers behind ships and planes, scientists have discovered 171 changes in polarity, or direction of the magnetic field, of the ocean bottom going back 76 million years.

Vine decided that this magnetic record could be highly useful. If the sea floor is actually spreading away from the mid-ocean ridges, he reasoned, the movement should be "seen" magnetically. For as the lava wells up from the ridges and hardens into crust, it is imprinted with the earth's magnetic field at the time. Furthermore, as pointed out in Chapter Four, the magnetic imprint will not normally change, even as the rock spreads away from each side of the ridge. Thus, a magnetic stripe of similar

78

RIDGE

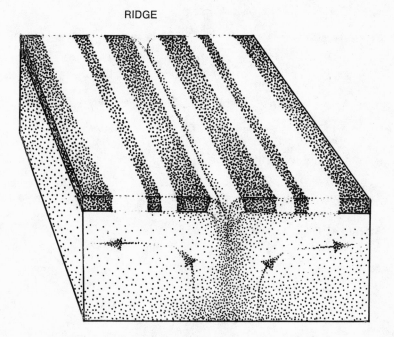

Corresponding magnetized stripes of lava at the sides of a ridge on the sea floor

polarity will run along each side of the ridge. Yet, when the earth's magnetic field changes, the new lava at the ridge will be differently imprinted and form a second pair of stripes of reversed polarity. In time, magnetic reversals should produce a whole series of alternating zebra-like stripes running parallel to the ridges. Vine himself compared the spreading ocean floor to magnetic recording tape, faithfully registering every switch in the earth's magnetic field as it reels out from the ridges. (Needless to say, these stripes cannot really be seen, except on the magnetic charts drawn by scientists from their readings.)

Vine's tape provided instant playback. Before he and his collaborator, Drummond H. Matthews of Cambridge, formally proposed their magnetic test of sea-floor

79

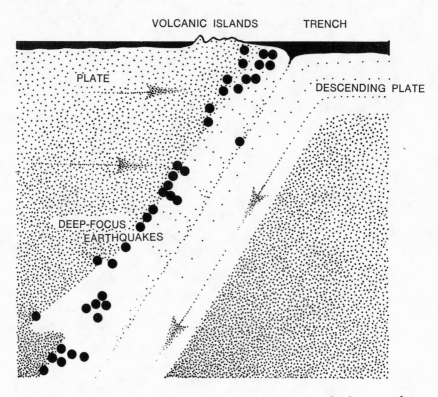

Deep-focus earthquakes mark the line at which two plates collide

spreading, considerable data had been accumulated. Some of the best evidence came from a magnetic survey of the Indian Ocean's Carlsberg Ridge (named in honor of the brewers of Denmark's Carlsberg beer for their lavish support of oceanographic research). There, the magnetic readings showed organized patterns of alternating stripes, running parallel to the ridge exactly as Vine had predicted.

A few years later, Vine and the Canadian geophysicist J. Tuzo Wilson provided still more support for sea-floor spreading. Using the known timetable of the

80

earth's magnetic reversals, they calculated the age of a number of individual magnetic stripes in selected parts of the ocean floor. What made this dating significant was that it offered a means of clocking the actual rate of sea-floor spreading.

If, for example, a particular stripe was found to be two million years old, that meant it had been traveling that long from its birthplace, the mid-ocean ridge. Then by dividing the age of the stripe by the distance it had covered in those two million years, Vine and Wilson could calculate the average annual speed of that part of the ocean floor. In one area near a ridge called the East Pacific Rise, the speed came to about two inches a year (other regions near this ridge move much faster); in the Atlantic, the speeds were somewhat slower. Nonetheless, the magnetic data indicated that the Atlantic had opened up 150 million to 200 million years ago. These figures were remarkably close to Wegener's own estimates half a century earlier for the breakup of Africa and North America.

The next important test of the Hess-Dietz hypothesis involved a different earth science. In the early 1960s, in order to keep watch for potential violations of the nuclear test ban treaty—which outlaws explosions of atomic bombs in the atmosphere and the seas—the United States had begun erecting a worldwide network of sensitive seismographs. These instruments could record earth tremors from secret nuclear explosions anywhere on the globe. For scientists, however, they served another function: they could pinpoint the origin of earthquakes with precision and also discern the different characteristics of their shock waves. As the data piled up from these new seismic listening posts, seismologists

81

began keeping worldwide maps, noting the occurrence of every tremor with an appropriately placed dot. Soon it became evident that the greatest number of dots were being placed in the vicinity of the mid-ocean ridges and the deep-sea trenches, especially those around the Pacific —a region aptly known as the "ring of fire."

These regions are also considered, in the theory of sea-floor spreading, to be sites of intense earth movements. Was there any connection between the theory and the seismic observations? Did sea-floor spreading have anything to do with the worldwide pattern of earthquakes?

Lynn R. Sykes and his fellow seismologists at Lamont thought the questions were significant enough to warrant serious study. If lava actually breaks through the earth's crust at the ridges, Sykes reasoned, these should be the site of shallow quakes, or earth movements close to the surface. Significantly, after examining the seismic records, Sykes found that the ridges were indeed the center of many shallow quakes. On the other hand, he thought that the deep-sea trenches should be the locale of deep-focus earthquakes because this is where the cooling lithosphere is pressing far down into the earth. This speculation was also confirmed by the earthquake records.

Nevertheless, Sykes was not quite ready to consider the case for sea-floor spreading seismically proved. He looked for one more piece of confirming evidence. Under the sea-floor-spreading theory, the lithosphere is presumed to be re-entering the earth at the trenches at an angle of about 45°. At the same time, the plate is thought to be thrusting under the neighboring lithosphere plate, coming at it from the opposite direction. If such head-on collisions occur in the trenches, Sykes thought, they should be easily detectable. Once more he sifted through the mounds of seismic data. And once more the search was productive.

Sykes found that most of the deep-focus quakes were originating along points that formed a slope of about 45°. This, of course, is the line along which the two plates should be meeting. Moreover, since the cooler, compacted lithosphere transmits seismic waves faster than the warmer, looser earth material around it, Sykes could trace the actual outline of the descending plate. As expected, its pattern conformed almost exactly to what had been predicted by the theory of sea-floor spreading.

The ease with which Hess's geopoetry passed these two distinctly different tests—one involving seismology, the other paleomagnetism—impressed many scientists. By the late 1960s, the critics of continental drift had become much less vocal and had dwindled in number. Still, there were important holdouts, especially among geologists, who are among the most conservative of scientists. For them, magnetic or seismic data would not suffice. Before they accepted the old Wegenerian dream, they would want undisputed physical evidence—which they could see, feel, and examine. This meant nothing less than actual rocks from the floor of the ocean.

CHAPTER EIGHT

The Challenger *Sails Again*

NOTHING is quite so simple as picking up a geological sample on dry land. At most it takes a hammer, a chisel, and a little muscle. Gathering a specimen from the bottom of the deep sea is another matter; the average depth of the ocean is more than two miles. Imagine trying to reach this far down from the deck of a ship bobbing fitfully in the waves.

Even if the ship could be held steady, the job is still an oceanographer's nightmare: the necessary cable alone would weigh several tons. The dredging tool would have to be strong enough to withstand the crushing pressures of the deep. Lowering the equipment would take hours of nerve-wracking effort, since any sudden lurch of the ship might snap the wire and send quantities of costly gear into the arms of Davy Jones.

Scientists, however, think the risks worth taking. The ocean floor is blanketed with layer upon layer of oozy sediments, made up largely of the skeletons of tiny meandering marine organisms called plankton (from the Greek for "wanderer"). These layers form an amazingly revealing record of the earth's past. By studying changes in the miniature fossils under microscopes, Lamont's David B. Ericson and Goesta Wollin have been able to chronicle the earth's periodic Ice Ages. One revealing clue was provided by a species of Foraminifera, the common plankton *Globorotalia menardii*. When the climate was cold, the shells of these animals spiraled to the left; when it was warm, they grew to the right. Similarly, the layers of the sediment can be used to date the ocean floor itself. By identifying fossils of known age in the lowest layer—which is also the oldest—scientists can accurately estimate when the volcanic bedrock underneath it was formed.

In their earliest attempts to collect deep-sea sediments, scientists scraped ooze off the bottom with primitive dredges or grabbing devices. But they soon realized that if they wanted to read these ancient records correctly they had to recover the layers in the order in which they were formed. The solution to that problem was a technique called coring: a pipe was forced as far as possible into the soft ocean floor, capturing a cylindrical sample of sediment inside the hollow tube. Then the pipe was hauled back aboard ship and the sediment was carefully extracted without disturbing the sequence of layers. The first cores, taken by the famous *Challenger* expedition, were only two feet long, yet their contents are still occasionally studied by scientists. Later, more efficient coring tools were developed. By using weights, suction, and sometimes explosives to drive in the tubes, oceanographers could sometimes obtain cores of 60 feet or more. But they obviously still had a long way to go. In typical areas of the oceans, soft sediments are piled up for 1,500 feet, on top of another 3,500 feet of harder rocklike sediment. Only by passing through these two layers could they reach the ocean floor's basaltic bedrock.

The alternative to coring is drilling. By the 1950s, oil companies had already drilled successfully from the decks of ships and barges, but these operations were rarely attempted in depths of more than 600 feet or beyond the continental shelf; the shelf is a sloping underwater extension of a continent rather than true ocean bottom. Nonetheless, American oceanographers were sufficiently impressed with this display of drilling skill to attempt a bolder assault on the ocean floor. This was Project Mohole, funded by the National Science Foundation.

Mohole's ambitious objective was nothing less than to penetrate the earth's mantle—an old dream of scientists and of science fiction, as in Jules Verne's *Journey to the Center of the Earth*. The project's name came from the so-called Moho or Mohorovicic discontinuity: the abrupt change in material that marks the boundary between the crust and mantle. It was discovered in 1909 by the Yugoslav geophysicist Andrija Mohorovicic after he analyzed earthquake shock waves. Yet locating the Moho was quite different from drilling into it. Though the mantle comes within three miles of the earth's surface under some parts of the ocean (compared with twenty miles or more under the thicker continental

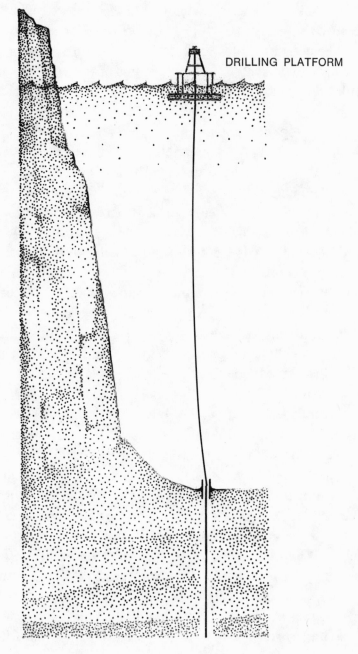

DRILLING PLATFORM

Project Mohole

crust), a sea drill would also have to go through two miles of water before reaching the Moho.

After a few test borings, the magnitude of Project Mohole's task became painfully apparent. The estimated cost soared to more than $100 million from only a fraction of that amount. Congressmen grumbled that Mohole was really a rathole for taxpayers' money. Many of the participants rightly charged mismanagement and serious quarreling broke out. Finally in 1966, much to the regret of its leading scientific advisers, including Princeton's Harry Hess—who had hoped to obtain new clues to the earth's origin, history and composition—Project Mohole was abruptly canceled by a chagrined Congress.

Though it had stopped thousands of feet short of its goal, Mohole was far from a flop. While working in the two-mile-deep waters off the coast of Mexico, Mohole's engineers had managed to overcome one of the greatest hurdles in ocean drilling: that of keeping a ship which is being buffeted by winds, currents, and waves over a drilling site. Anchoring in the open sea would have been futile; if the anchor chain did not break under its own weight, it would still have been too slack to hold the ship on station. So Mohole's engineers ingeniously attached four large diesel-powered outboard engines to the hull of their drill ship, locating two engines opposite each other at each end of the vessel. The engines also could be individually controlled. Thus, the helmsman could use them singly or in combination, at whatever power necessary, to halt drifting of the ship caused by wind or seas coming from any direction. Called "dynamic positioning," this new technique enabled drillers to hold a ship in place with its engines alone.

Encouraged by the technological breakthrough, the   89

Scripps Institution of Oceanography of the University of California in San Diego and four other leading United States ocean research centers launched another attack on the ocean floor. In addition to Scripps, members of the group, known as Joint Oceanographic Institutions for Deep Earth Sampling (JOIDES), were Columbia University's Lamont-Doherty Geological Observatory, the Institute of Marine Sciences of the University of Miami, the University of Washington (Seattle), and the Woods Hole Oceanographic Institution. Their program was called the Deep Sea Drilling Project (DSDP) and was also financed by the National Science Foundation. Less ambitious and dramatic than Mohole, DSDP had no intention of sinking one long hole through the crust. Instead, its objective was to drill many shallow holes through the sediments on top of the ocean bedrock.

This tactic not only increased the likelihood of successful drilling, since it would not be necessary to attempt the difficult job of replacing worn-out drills at sea; it also had certain scientific advantages. In the eyes of some oceanographers, fossil-rich sediments are often a better record of the earth's past than the underlying bedrock. Furthermore, by spending only a short time at one site, the drillers would be able to take sediment cores from all the major seas. Thus, DSDP would be able to accomplish one of its major objectives: to collect enough material from widely scattered parts of the ocean floor for a decisive test of the theories of sea-floor spreading and continental drift.

For that globe-girdling assignment, a special ship had to be designed and built. It was christened *Glomar Challenger* in honor of the pioneering British oceanographic vessel of a century before (Glomar is a contraction of the name of the ship's owners, Global Marine Inc., from whom it was chartered by Scripps). In other respects, the new *Challenger* has little in common with its canvas-rigged predecessor. Topped by a 142-foot-high drilling tower amidships and stacked with miles of drill-pipe sections on its decks, the vessel looks like flotsam from an offshore oil field.

In spite of its ungainly appearance, the *Challenger* is one of the most unusual research vessels afloat. It has six separate laboratories for studying deep-sea cores. Its maze of electronic gear includes computers, underwater listening devices, and even radios to pick up navigational information and weather pictures from satellites orbiting far overhead. Its most extraordinary equipment, how-ever, is far below deck. Besides the twin main screws at the stern, the *Challenger* has two pairs of smaller propellers, or thrusters, located fore and aft in tunnels through the sides of its hull. Like the experimental Mohole ships, the *Challenger* uses these props for dy-namic positioning over a drill site far out at sea. They are, in fact, so effective that the ship can maintain its station in battering waves twelve feet high. The *Chal-lenger* is also equipped with gyroscopically controlled balance tanks that sharply reduce pitching and rolling, and also avoid seasickness among the seventy scientists and crew members.

Life aboard the *Challenger* during a typical drilling operation is a scientific adventure. The ship is far out at sea, performing with its usual robot-like efficiency. Once it has been positioned over the drill site and everything is ascertained to be in order, the captain gives the com-mand "Make hole!" A small box containing the acous-tical signal beacon is dropped overside. On the bridge, the ship's electronic listening devices are switched on. The computer is told to take over control of the propellers. Having been fed the precise location of the intended drill site, the electronic brain knows exactly what to do. It will sense any drifting by changes in the angle of the sound waves coming from the acoustical beacon on the ocean floor. Then, to bring the *Challenger* back on station, it will activate one or more of the thrusters and, if necessary, the main propellers. Through the electronic wizardry, the ship is, in effect, anchored to the bottom by invisible lines of sound.

Now the drilling crew swings into action. Many of them are roustabouts from the offshore oil rings of the Gulf of Mexico. But since much of the pipe-handling is automated, there is little need for brute strength. It is their job to piece together the "drill string"—the series of linked pipe sections that will reach down to the ocean floor. Lowered through the 20-by-22-foot well in the ship under the drilling tower, the first section to dip into the water is the heavy drilling pipe and outer core barrel; it is tipped with a large diamond-studded or tungsten bit, the drilling tool that will actually do the cutting. More heavy pipes follow, because the lower end of the drill string must have enough weight to press hard against the ocean floor, if the bit is to drill successfully. To cushion the drill string against the ship's up-and-down motion, the pipes are connected by sliding joints, like shock absorbers on a car.

Next, the crewmen begin attaching lighter pipes, but they must still show extreme care. In spite of the ship's stability, an unexpected roll or heave can put a severe strain on the topmost part of the drill string. Once

The Glomar Challenger *at work in the Gulf of Mexico*

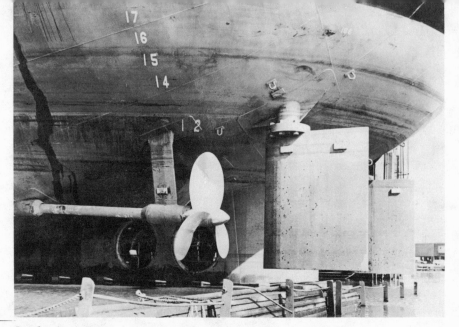

In drydock, main propellers, rudders, and thrusters (openings behind propeller) are inspected

The crew prepares to add a section to the drill string

Opposite: *Rigging the machinery that turns the drill string*

OPPOSITE: *Melvin M. A. Peterson and N. Terence Edgar examine a freshly drilled core*

Peterson and other officials inspect a new tungsten drill bit

The re-entry cone goes overboard

14,000 feet of pipe was lost when a sudden jolt snapped the drill string at a point just under the hull.

Throughout, the crew works with confidence. After four hours of locking and lowering pipes from the drill tower, 11,000 feet of drill string dangles beneath the *Challenger*. Suddenly the ship's instruments register a reduction in the weight of the piping; the drill bit has touched bottom. The drilling superintendent orders his men to attach the power unit and commence drilling, and the long strand of pipe slowly begins to revolve, twisting like a monstrous snake. At its other end the bit starts cutting into the ocean floor. The *Challenger*'s scientists are delighted. As the drilling proceeds smoothly, it becomes clear that they have chosen a good site.

When the bit has penetrated far enough, the scientists ask to take their first core. The drilling is temporarily stopped while a thin wire running through the drill string pulls up the removable centerpiece of the drill bit. A 2½-inch-wide core tube is sent down in its place, and the drilling is resumed until the thin pipe is filled with sediment. Then the tube is hauled up again, and the core, nearly thirty feet long, is carefully removed. After a quick inspection, the scientists decide against any further coring at that depth and order the drillers to penetrate deeper.

While the drill churns away, the scientists deftly cut the precious core into more manageable sections. Next, they split the sections lengthwise; half are put into a refrigerated van where they are preserved in their original condition for examination by researchers on shore; the other half are at once analyzed in the *Challenger's* laboratories. By the time the scientists are done with photographing, x-raying, pounding and slicing,

99

radiation tests, mineral identification, and fossil inspections of the core, its characteristics seem as familiar as those of an old friend.

Many hours later, after the drill bit has reached maximum depth and the last core has been removed, the crew begins the laborious retrieval of the drill string. Reversing the original operation, the lift operators haul back one section of pipe after another until the two miles of tubing is back on deck. Exactly as much concentration as before is required from the *Challenger*'s men to prevent damage or loss of valuable pipe. As it is stowed away, the pipe is inspected and treated for corrosion, since a flaw in one section might cause loss of a whole drill string in future drilling. Only one thing is left—intentionally—on the ocean floor: the inexpensive signal beacon.

Since DSDP's first drilling leg in the summer of 1968, the *Challenger* has made oceanographic history time and again. In some places, the drill string has stretched nearly four miles under the ship. The *Challenger* has also sunk holes of more than half a mile into the ocean floor. But even these records probably will not stand for long.

In the past, a hole had to be abandoned once the drill bit became dull and could not cut. With the development of a so-called re-entry system, the *Challenger* can replace worn-out bits and return the drill string to the same hole. The trick is to leave behind a large funnel over the hole after the drill string is removed. Equipped with sonar beacons, the funnel gives off signals that serve as guides for the drill string's return to the hole. The drill barrel itself is equipped with a remote-controlled jet or thruster that lets the men aboard ship maneuver the drill string into the funnel—and the hole—to

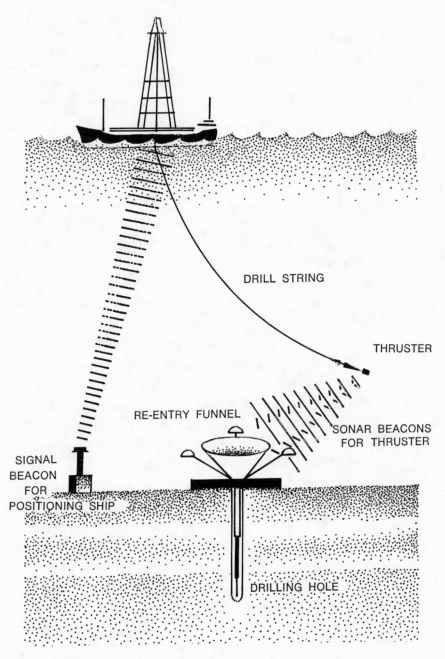

DRILL STRING

THRUSTER

RE-ENTRY FUNNEL

SONAR BEACONS
FOR THRUSTER

SIGNAL
BEACON
FOR
POSITIONING SHIP

DRILLING HOLE

The re-entry system of the Glomar Challenger

resume drilling. The feat involves extraordinary marksmanship. A Scripps scientist says: "It's like dangling a wire from the top of the Empire State Building and dropping its other end into the neck of a Coke bottle on the street below." With this exceptional re-entry system, the *Challenger* may eventually accomplish what Project Mohole never could: drilling into the mantle.

The *Challenger*'s achievements have already exceeded expectations. On its very first voyage into the Gulf of Mexico, it discovered oil and natural gas deposits far out at sea; the drill hole was prudently capped with cement to prevent leakage. It has also found hints of how and where minerals such as manganese are formed. Most important of all, the *Challenger* has produced what the former chief scientist of DSDP, Dr. Melvin N. A. Peterson of Scripps, calls almost unimpeachable evidence for sea-floor spreading and continental drift.

The sediments, for one thing, were older and thicker as the ship drilled farther from the mid-ocean ridges. Moreover, the rate of spreading calculated from the age of the cores was uncannily close to the forecasts. The North Atlantic, for example, appears to have opened up some 170 million to 180 million years ago—with Europe and North America separating at an average rate of an inch a year. Even more impressive, the rate of spreading sometimes greatly surpassed predictions. In an equatorial area of the East Pacific Rise, for instance, the ocean floor seemed to be growing at the phenomenal rate of up to twelve inches a year. According to Peterson, "During the lifetime of a man, the sea floor can move a distance easily the length of his body, and the rapid rates approach those at which common trees grow."

These voyages are only a beginning, just as was the

expedition of the *Challenger*'s namesake a century ago. In the future, scientists will probe deeper into the ocean floor, uncovering more of its secrets. But by providing the first physical evidence of continental drift from the sea bottom, the *Glomar Challenger* has already assured itself a proud place in the annals of science. Never again will skeptics be able to say that there is no solid geological evidence of continental drift.

CHAPTER NINE

The Ever-Changing Earth

H OW simplified scientists' problems would be if they could turn back the clock as did H. G. Wells in his story *The Time Machine*. Then, with only a flick of a switch, earth scientists could travel far back into the earth's dim past and observe for themselves the shape and location of the continents. In a sense, the geologists Robert Dietz and John S. Holden have gone on such an improbable journey. Their time machine was a modern computer. With its help, they analyzed the mass of new evidence in favor of continental drift, pieced together the continents into the best possible fit, and then traced their gradual movements apart over the slow passage of geological time. 105

Their electronic journey took them far back into the Triassic period, more than 200 million years ago, to the dawn of the age of dinosaurs. Most of the familiar terrestrial creatures, including man, had not yet appeared. The face of the earth itself was barely recognizable. Instead of seven continents, Dietz and Holden found only one: a huge supercontinent strikingly similar to Wegener's own Pangaea. This land mass, in turn, was surrounded by a single universal sea, like Wegener's Panthalassa.

Pangaea, in the computerized reconstruction, covered about 40 percent of the earth's surface. Had there been a country called the United States at the time, it would have been located far to the southeast of its present position; New York would have been near the equator. In contrast, the land that was to become Japan, now roughly in the same latitudes as California, was far to the north, near the Arctic Circle. India and Australia were located near the opposite end of the world, tightly packed against Antarctica.

Looking over this supercontinent, Dietz and Holden could not tell how long Pangaea had existed or whether there had been any continents before it. But it was clear to them that its present life was now coming to an end. Large flows of volcanic material were welling up slowly from the earth's interior along two great cracks in the middle of what would become the North Atlantic and Indian oceans. Under the pressure of this lava, Pangaea was gradually, almost imperceptibly tearing apart.

With the help of their electronic time machine, Dietz and Holden leaped forward across another 20 million years of earthly history. That brought them to the start of the Jurassic period, 180 million years ago.

In the northern half of the globe, Pangaea had given birth to a smaller continent known to geologists as Laurasia: the ancestor of North America, Europe, and Asia. In the southern half, Pangaea's breakup had led to the formation of a second continent. It was Eduard Suess's old Gondwanaland, the parent continent of Africa, India, South America, and Antarctica. In fact, there were signs that Gondwanaland itself was not long for the world. India, breaking free from the rest of Gondwanaland along a Y-shaped rift, had managed to separate one half of the remaining continent (South America and Africa) from the other (Australia and Antarctica). Laurasia, meanwhile, was undergoing a less violent change: it was rotating slightly clockwise, thereby enlarging the newborn North Atlantic.

Continuing their imaginary journey, Dietz and Holden arrived in the early Cretaceous period, 135 million years ago, when the first true birds were appearing in the primeval skies. On the ground, the geological drama was continuing. South America was breaking away from one side of Africa, the island of Madagascar from the other side. Meanwhile, Africa itself was drifting slowly northward; this passage continued for millions of years until Africa finally brushed up against Asia and closed off the eastern end of the Mediterranean Sea. Simultaneously, the North Atlantic was growing larger as North America moved farther and farther from Europe. In the process, a second rift developed in the Atlantic, beginning the formation of Greenland.

Then, taking another jump in time, Dietz and Holden entered the most recent geological era, the Cenozoic, which began 65 million years ago. By now the earth was covered with many familiar trees and   107

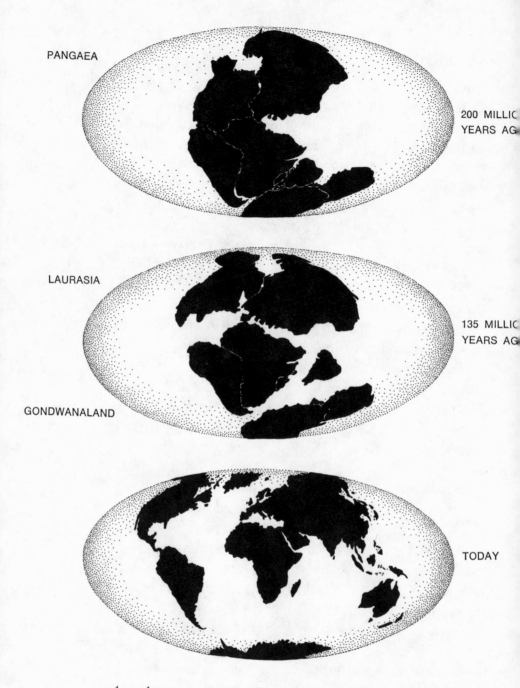

PANGAEA

200 MILLIC
YEARS AG

LAURASIA

135 MILLIC
YEARS AG

GONDWANALAND

TODAY

A modern conception of the way Pangaea broke up

plants; some mammals had also appeared, although man's arrival was still millions of years away. The continents, too, were taking on a more familiar look. The two Americas, though born separately, were finally united by the Isthmus of Panama, a small strip of land created largely by volcanic eruptions. India, after traveling thousands of miles north, crashed into the underbelly of Asia, crumpling so much crust in the collision that the Himalayas, the world's highest mountains, were left as wreckage. In fact, the continents had scattered so far afield that there was hardly a hint—except the telltale fit—of the existence of Pangaea.

Was Dietz's and Holden's computerized journey into the past a flight of fancy? Or was it an accurate reconstruction of the breakup of Pangaea and the drift of the continents? Surprising as it may seem, there are still a few scientists, notably the Russian geophysicist Vladimir V. Beloussov, who have doubts about continental drift. In particular, they question the idea that there are currents in the earth's interior powerful enough to propel the continents. This motive force also troubles some believers in drift, and they have suggested that supplementary mechanisms, including gravity, may help move the continents. But the overwhelming majority of earth scientists accept continental drift as a geological certainty. So convincing is the evidence, in their eyes, that it has given rise to an entirely new way of looking at the earth, called "plate tectonics." (Tectonics is the word for any process by which the earth's surface is reshaped.)

Under plate tectonics, a theory largely developed by the Princeton geophysicist W. Jason Morgan, the earth's outer shell is no longer regarded as rigid and fixed. In-　109 stead, it is thought to be made up of about half a dozen huge plates, each about seventy miles thick. They are composed of the earth's thin outermost crust and the harder, underlying lithosphere; their boundaries are the mid-ocean ridges and the deep-sea trenches, where (as explained in Chapter Six) they add or lose material. In addition, the major plates are themselves divided into smaller subplates.

As they float on top of the mantle's asthenosphere, the plates may bring about great upheavals of the earth's surface. When a plate tears apart, for example, a new continent may be born; Europe and North America were apparently created when the plate underneath Laurasia broke apart at the Mid-Atlantic Ridge. When two plates collide, on the other hand, the thrust of one under the other may uplift enough crust to build high mountains, like the Andes and Himalayas, or create long chains of volcanic islands, like Japan and the Philippines. And, of course, the continents themselves are carried by the plates as they move over the face of the earth.

Normally, such plate or tectonic activity is extremely slow, but occasionally it may be rapid—and dangerous. If two plates happen to be sliding past each other in opposite directions, as are the Pacific and North American plates under California's San Andreas fault, they sometimes become stuck. This causes the buildup of tremendous strains in the earth, which must inevitably be released. When they are, the plates lurch forward, like overwound clock springs. Though these jolting movements may measure no more than a few feet and last only a few seconds, they are often powerful enough to be observed as violent tremors or earthquakes.

110   Why has the earth's surface broken up into plates?

How long have these plates existed? No one really knows the answers to these terrestrial mysteries, but scientists suspect that the existence of the plates predates even the breakup of Pangaea. One clue to such ancient tectonic activity is a mountain range—the Urals—running north and south through the center of the Soviet Union. If it is true that mountains are formed at a continent's edge by the collision of plates, why then are the Urals so far inland?

Attempting to answer that question, some scientists have suggested that the Urals were created by the collision of the plates that may have formed the Laurasia portion of Pangaea. If their intriguing theory is true, it would have profound implications: it would mean that the continents were breaking apart, drifting, and coming together again long before the appearance of Pangaea.

Such speculations only emphasize the promise of the revolution in geological thinking brought about by the acceptance of continental drift. In the not too distant future, the discoveries of recent years may well be utilized for man's benefit. Scientists are talking about using their new knowledge in the search for vital minerals. Above the Red Sea rift, for instance, oceanographers have discovered hot, mineral-rich waters, laden with gold, zinc, copper, and silver; someday these ores may be profitably extracted from the sea.

The new geology may also lead to the discovery of valuable sources of energy. Iceland, which is situated on top of the geologically active Mid-Atlantic Ridge, makes much of its electrical power with the steam that pours out of hot springs, or geysers. So do parts of California, Italy, and New Zealand. With their new knowledge of the earth, scientists may be able to find more

111

sources of such cheap, pollution-free geothermal energy. The environment may also be helped in other ways by the new geology; some ingenious scientists have suggested using the deep-sea trenches, where the plates are being drawn back into the earth, as garbage dumps for man's growing piles of waste. At the very least, the new geology should teach man how to predict and perhaps to tame earthquakes. For it is a truism of science that understanding a process points the way to altering or controlling it.

Most important of all, the new geology has given man a totally new view of his home planet. No longer can he think of the earth as a dull, unchanging place. Instead, he knows that it is a restless, tireless old globe whose face, as Wegener so perceptively realized, has

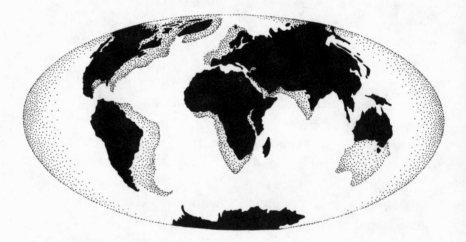

The way the earth may look 50 million years from now (Dotted areas represent the present shape and position of the continents)

been remade in the past and will change again in the future.

What shape will that future take?

No one can really tell, but Dietz and Holden have turned to their electronic time machine for an answer. By projecting current plate movements into the future, this is the picture they have drawn up of what the earth might look like 50 million years from now:

The Atlantic has expanded at the cost of the Pacific, which has slightly decreased in size. Australia has drifted northward, but will avoid what looks like a certain collision with Asia. Along the east coast of Africa, a giant chunk of land has been torn free. The two Americas have broken apart, ending the need for the Panama Canal. So have Africa and Asia, thus putting the Suez Canal out of business, too. Similarly, the sliver of California west of the San Andreas fault has split off from the rest of North America and pressed northward. Located on the sliver, Los Angeles has long since passed by San Francisco, which has remained attached to the mainland side of the fault. In fact, this piece of America has begun to plunge into the Aleutian trench, where it will be consumed by the mantle.

Where does it all end? Not even Dietz and Holden dare to say.

SELECTED BIBLIOGRAPHY

INDEX

SELECTED BIBLIOGRAPHY

BOOKS

Barton, Robert. *Oceanology Today*. Garden City, N.Y.: Double-
day, 1970.
Description of the tools and techniques of oceanography
(oceanology), including a chapter on the *Glomar Challenger*.
Bascom, Willard. *A Hole in the Bottom of the Sea*. Garden City,
N.Y.: Doubleday, 1961.
Account of the origins of Project Mohole by one of the men who
helped begin it.
Behrman, Daniel. *The New World of the Oceans*. Boston: Little,
Brown, 1969.
Readable account of the work under way at leading oceano-
graphic institutions.
Beiser, Arthur. *The Earth*. *Life* Nature Library, new ed. New York:
Time-Life Books, 1970.
Includes a well-illustrated section on the new earth theories.
Cailleux, André. *Anatomy of the Earth*. New York: McGraw-Hill
Book Company. 1968.
Basic primer on the earth by a French scientist. For high school
and college students.

Clayton, Keith. *The Crust of the Earth*. Garden City, N.Y.: Natural History Press, 1967.
Well-illustrated explanation of geological history and theory. For younger readers.

Cowen, Robert C. *Frontiers of the Sea*. Rev. ed. Garden City, N.Y.: Doubleday, 1969.
The story of oceanographic exploration—including the original *Challenger* voyage—by the science editor of *The Christian Science Monitor*.

Ericson, David B., and Wollin, Goesta. *The Ever-Changing Sea*. New York: Alfred A. Knopf, 1967.
History of the exploration of the sea floor by two skilled workers in the field.

Longwell, Chester R.; Flint, Richard Foster; and Sanders, John E. *Physical Geology*. New York: Wiley, 1969.
Basic text, including many of the new theories. For advanced high school and beginning college students.

Moore, Ruth. *The Earth We Live On*. Rev. ed. New York: Alfred A. Knopf, 1971.
A lively introduction to geology for the general reader; the new edition has been updated to include the latest discoveries.

Nature/Science Annual. 1970 Edition. New York: Time-Life Books.
The chapter entitled "The Wandering Continents" by Tom Alexander summarizes the case for continental drift in a clear, nontechnical way.

Phillips, O. M. *The Heart of the Earth*. San Francisco: Freeman, Cooper, 1968.
A look at the earth by a practicing geophysicist, including a description of terrestrial magnetism. College level.

Runcorn, S. K. (ed.). *Continental Drift*. New York: Academic Press, 1962.
Papers on various aspects of the subject, some technical; a biographical reminiscence of Alfred Wegener will appeal to the general reader.

Takeuchi, H.; Uyeda, S.; and Kanameri, H. *Debate about the Earth*. Rev. ed. San Francisco: Freeman, Cooper, 1970.
Review of the case for continental drift by three Japanese earth scientists. College level.

Wegener, Alfred. *The Origins of Continents and Oceans*. 4th ed. Translated by John Biram. London: Methuen, 1967.
This translation of the final (1929) German edition of Wegener's classic is worth exploring by serious readers. Contains an introduction by the English geologist B. C. King and a brief biography of Wegener by his brother, Kurt Wegener.

ARTICLES

Andel, Tjeerd H. van. "Deep-Sea Drilling for Scientific Purposes: A Decade of Dreams," *Science*, June 28, 1968, pp. 149-1424.
The reasons for sending the *Glomar Challenger* to sea.

Craddock, Campbell, "Antarctic Geology and Gondwanaland," *Science and Public Affairs*, December, 1970, pp. 33-39.
The importance of the geology of Antarctica in understanding the forces at work in the earth.

Dietz, Robert S., and Holden, John C., "The Breakup of Pangaea," *Scientific American*, October, 1970, pp. 30-41.
Traces the break-up of Wegener's supercontinent with the help of a computer.

Heirtzler, J. R., "Sea-Floor Spreading," *Scientific American*, December 1968, pp. 60-70.
How the new earth theories may account for such phenomena as volcanoes and earthquakes.

Horsfield, Brenda, and Dewey, John, "How Continents Are Made and Moved," *Science Journal*, January, 1971, pp. 43-48.
Explanation of plate tectonics.

Hurley, Patrick M., "The Confirmation of Continental Drift," *Scientific American,* April 1968, pp. 53-64.
A summary of the evidence.

Pitman, W. C., "Sea-Floor Spreading," *Science Journal*, February 1969, pp. 51-56.
How the case was proved.

Rupke, N. A., "Continental Drift Before 1900," *Nature*, July 25, 1970, pp. 349-350.
Clears up many misunderstandings about the historical origins of the idea.

Tazieff, Haroun, "The Afar Triangle," *Scientific American*, February 1970, pp. 32-40.
Theory that the Red Sea, located at the juncture of three great rifts, is actually an ocean in the making. By a well-known vulcanologist.

Vine, F. J., "The Geophysical Year," *Nature*, September 5, 1970, pp. 1013-1017.
Tells how the theory of sea-floor spreading has been updated by the new global tectonics.

Wilson, J. Tuzo, "Continental Drift," *Scientific American*, April 1, 1963, pp. 86-100.
One of the earliest comprehensive articles on the new scientific interest in the subject.

INDEX

121